Kids Nonfiction - PORTR-KOU
304.28 KERRY
Kerry, Isaac
Climate change migrants
33410018298770   01-12-2023

Spotlight on Climate Change

Climate Change Migrants

Isaac Kerry

Lerner Publications ◆ Minneapolis

To Lily and Julia

Copyright © 2023 by Lerner Publishing Group, Inc.

All rights reserved. International copyright secured. No part of this book may be reproduced, stored in a retrieval system, or transmitted in any form or by any means—electronic, mechanical, photocopying, recording, or otherwise—without the prior written permission of Lerner Publishing Group, Inc., except for the inclusion of brief quotations in an acknowledged review.

Lerner Publications Company
An imprint of Lerner Publishing Group, Inc.
241 First Avenue North
Minneapolis, MN 55401 USA

For reading levels and more information, look up this title at www.lernerbooks.com.

Main body text set in Adrianna Regular.
Typeface provided by Chank.

Editor: Brianna Kaiser

Library of Congress Cataloging-in-Publication Data

Names: Kerry, Isaac, author.
Title: Climate change migrants / Isaac Kerry.
Description: Minneapolis : Lerner Publications, [2023] | Series: Searchlight books - spotlight on climate change | Includes bibliographical references and index. | Audience: Ages 8–11 | Audience: Grades 4–6 | Summary: "With increasing environmental changes resulting from climate change, many people have had to move away from their homes. Readers will learn about the relationship between climate change and migration in this up-to-date title"— Provided by publisher.
Identifiers: LCCN 2021053762 (print) | LCCN 2021053763 (ebook) | ISBN 9781728457949 (library binding) | ISBN 9781728463933 (paperback) | ISBN 9781728461915 (ebook)
Subjects: LCSH: Climatic changes—Juvenile literature. | Forced migration—Environmental aspects—Juvenile literature.
Classification: LCC QC903.15 .K47 2023 (print) | LCC QC903.15 (ebook) | DDC 304.2/8—dc23/eng/20211223

LC record available at https://lccn.loc.gov/2021053762
LC ebook record available at https://lccn.loc.gov/2021053763

Manufactured in the United States of America
1-50822-50161-2/17/2022

Table of Contents

Chapter 1
LIVING IN A WARMER WORLD . . . 4

Chapter 2
CHANGING CLIMATE, CHANGING LIVES . . . 12

Chapter 3
THE AGE OF CLIMATE MIGRATION . . . 20

Chapter 4
FINDING SOLUTIONS . . . 24

You Can Help! • 29
Glossary • 30
Learn More • 31
Index • 32

Chapter 1

LIVING IN A WARMER WORLD

Rising oceans. Raging storms. Scorching heat. Climate change is happening and impacts millions of people all over the world. As Earth warms and its climate keeps changing, more people will be affected. Certain areas of the world will become much harder to live in. Millions of people will have to make a choice: Will they stay where they are or seek a new life somewhere else?

Past, Present, and Future Climate Change

Scientists have collected data about Earth's surface temperatures since 1880. This gives them over 140 years of information about climate to look at and study. The data shows that the world is getting warmer, and it is warming quickly.

Scientists study climate trends to predict if tornadoes and other weather extremes are more likely to occur in the future.

Earth's hottest recorded decade took place between 2010 and 2019. The years 2019 and 2020 were among the top three warmest years ever. Every three years, a record is being set for the highest temperature.

Scientists can look at information about past climate patterns to help them understand how the climate is changing. To predict how Earth's climate will change in the future, they use climate models. These are a type of computer program. Using math and science, scientists test how factors such as population size and use of fossil fuels can affect climate change over time.

Scientist Ian Bartholomew measures how Greenland's Russell Glacier is impacted by climate change.

A HOT DAY IN BELGIUM IN 2020, ONE OF THE WARMEST RECORDED YEARS

Even a small increase in temperature can make a big difference. Weather patterns, water supplies, and natural ecosystems all exist in a delicate balance. Minor changes to temperature can create major shifts in these systems. When these systems change, environments can become harder to live in.

STEM Spotlight

Human activity is the main reason Earth is warming. Fossil fuels, such as oil and coal, are used to create energy. Businesses, cars, and homes all use fossil fuels for power. Burning these fuels releases greenhouse gases. These gases trap heat in the atmosphere and cause Earth's temperature to rise.

Renewable energy sources don't produce greenhouse gases. These energy sources, such as solar and wind, are better for the environment. Using more renewable sources will reduce the amount of fossil fuels used.

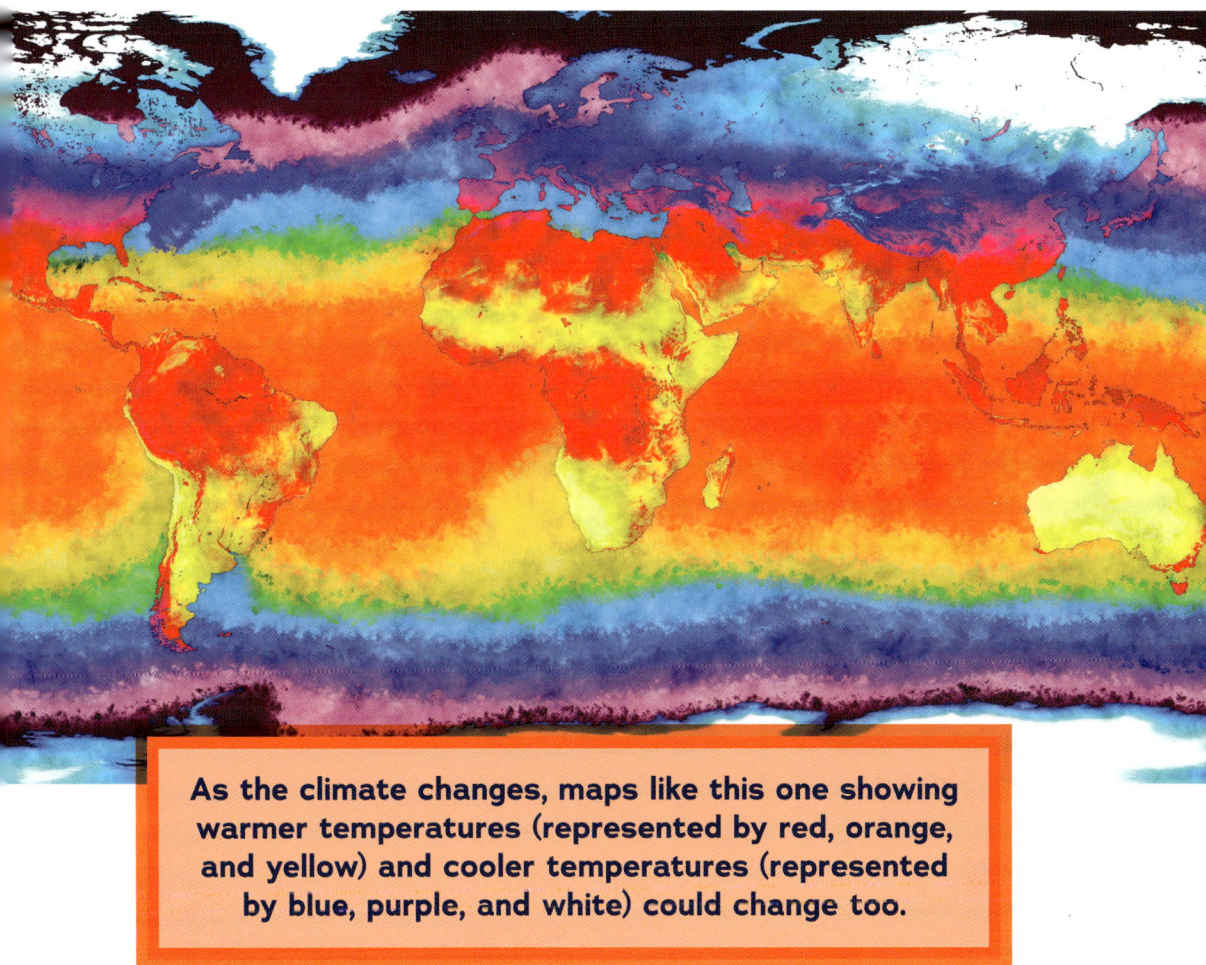

As the climate changes, maps like this one showing warmer temperatures (represented by red, orange, and yellow) and cooler temperatures (represented by blue, purple, and white) could change too.

Climate Migration

Many kinds of migration occur all over the world every day. About 3.5 percent of the world's population, or 272 million people, are international migrants. They live in a country other than their birthplace. People choose to move for a number of reasons.

Climate migration is a specific type of migration. Climate migrants are people or a group of people who must leave where they live due to changes in the environment caused by climate change.

Rising sea levels can impact people living on coastlines all around the world.

Climate change makes the conditions for hurricanes more common.

Climate change does not impact the world equally. Some places will get much warmer than others. Many of the countries that will be hit the hardest are polluting much less than other countries. So the countries that will be impacted the most have done the least to cause the problem. Even within other countries, the effects will not be felt equally. Some people will have the means to be able to move away, but others will have a much harder time leaving and be forced to adapt.

Chapter 2

CHANGING CLIMATE, CHANGING LIVES

As Earth's climate changes, dangerous weather events become more likely to happen. These affect millions of people around the world each year, and this number can grow with further climate change. People may have to leave their homes and move to other places because of the threat of these events. These events include heat waves, droughts, wildfires, and more.

Heat Waves

Warmer temperatures each year will mean more heat waves. A heat wave is a period of especially high heat. They can last a few days or go on for weeks. Climate change will make heat waves hotter and last longer. Heat waves are especially serious for cities. Buildings and roads of a city absorb more heat than places outside cities with more trees, grass, and plants do. This is the heat island effect.

A WOMAN IN BRAZIL COVERS HER HEAD FROM THE SUN DURING A HEAT WAVE.

STEM Spotlight

As the world heats up, many people will likely use air-conditioning more. This will use more energy and create more greenhouse gas emissions. A team of researchers is trying to solve this problem with something surprising: paint. Scientists at Purdue University have created the whitest paint in the world. This paint reflects over 98 percent of sunlight that hits it. The team estimates that painting a roof with their paint will create as much cooling as running an air conditioner.

Droughts and Wildfires

Droughts happen when there is little to no rain for a long time. Changing weather patterns will make certain areas of the world get much less rain. This creates a water shortage. Then there is not enough water for everyone. Higher temperatures make droughts worse because water in the ground evaporates. This makes areas drier. Droughts can also lead to food shortages.

Heat waves and droughts can damage crops.

More droughts mean more dry areas, which can lead to more wildfires. When the ground is dry, it is easier for fires to start. In recent years, this has been a big problem in places such as California and Australia.

Extreme Storms and Rising Seas

When hotter temperatures make more water in the ground evaporate, more water vapor is put in the air. The combined temperature and vapor in the air allow for dangerous storms such as hurricanes to form. And oceans are warming. A warmer ocean makes hurricanes more likely. These storms can also lead to flooding and destroy homes.

All these factors can lead to climate migration since people may move to avoid the threat of storms.

Hurricane Irma brings strong winds and water to the streets of Florida in 2017.

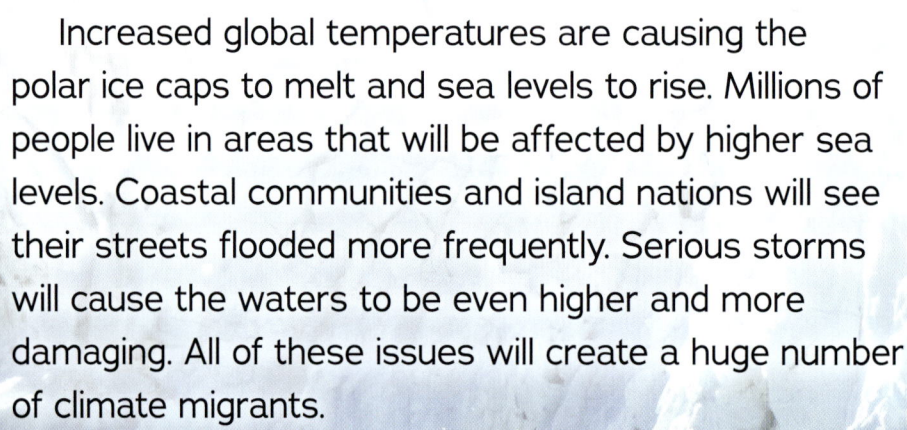

Increased global temperatures are causing the polar ice caps to melt and sea levels to rise. Millions of people live in areas that will be affected by higher sea levels. Coastal communities and island nations will see their streets flooded more frequently. Serious storms will cause the waters to be even higher and more damaging. All of these issues will create a huge number of climate migrants.

Part of a glacier breaks away in Alaska.

Pacific Climate Warriors

Pacific Climate Warriors is a group of 350 volunteers who are part of the global climate change movement 350.org. Pacific Climate Warriors are from Pacific island nations and use nonviolent ways to protest the fossil fuel industry. In 2014 they blocked coal ships by paddling handmade canoes. They also raise awareness and teach others about the effects climate change has on Pacific islands. Pacific Climate Warriors and other groups are taking a stand to protect their homes.

Chapter 3

THE AGE OF CLIMATE MIGRATION

Many people are already climate migrants. This number is likely to increase with further climate change. Some people will be able to migrate internally by moving to a new place in their home country. Other people will leave their country completely. This is called external migration.

Global Climate Migration

Some areas are more likely to be affected by climate change than others. That will make those areas harder to live in. It is estimated that by 2050 over 143 million people from Latin America, parts of Africa, and Southeast Asia will become climate migrants.

People from every part of the world will be affected by climate change and could become climate migrants.

A flooded neighborhood in Texas

In the United States many people will be forced to move away from the coasts since storms and higher seas will make living there too difficult. Other parts of the country, especially the Southwest, will become much hotter. The heat combined with drought conditions may cause people to flee.

Even countries that are not as impacted by climate change will be affected by climate migration. Many of the climate migrants seeking a new home will move to these countries.

PEOPLE IN AFGHANISTAN FILL UP WATER CONTAINERS FOR MIGRANTS IN 2019.

Chapter 4

FINDING SOLUTIONS

People and governments can help prevent further climate change. They can try to reduce global warming. Each additional degree of warming makes extreme weather more likely and climate migration more common.

On December 12, 2015, many countries from around the world adopted the Paris Agreement. Its goal is to limit global warming to below 2.7°F (1.5°C). Countries work toward this goal by reducing their greenhouse gas emissions. Turning to renewable energy sources will help reduce the amount of those emissions. More work is needed to meet this goal.

High school and college students hold a climate change protest in Belgium in 2019.

Hindou Oumarou Ibrahim

Hindou Oumarou Ibrahim, shown below in 2014, is from Chad, a country in Africa. She is a member of the Mbororo (also known as Wodaabe), a group of Indigenous people. When she was young, other children made fun of her for being a Mbororo. This made her aware of the discrimination her people faced. As she grew up, she realized the effects of climate change were another form of discrimination. As an activist, she fights for the rights of Indigenous people, women, and the environment.

A 2019 youth climate change protest in London, England

World governments must also prepare for further climate migration. Countries must plan and create policies designed to help migrants. By having a plan, governments can help ease migrants into their new homes.

People help by becoming more aware of climate change and spreading the word that it is a serious problem. They can also write to members of governments and companies to hold them accountable for their contributions to climate change. People can also do things every day to lower their impact on the

environment. Driving electric cars or carpooling, turning off lights and electronics when they are not being used, and recycling can all help the planet.

Some amount of climate change is going to happen, and this will lead to climate migration. By taking action, we can limit global warming and prevent millions of people from having to leave their homes.

Using renewable energy, such as wind energy from turbines, helps the environment.

You Can Help!

An important thing you can do is to let the people who represent you in government know you care about climate change. By making your voice heard, you can encourage our leaders to take these problems seriously. Share your opinion, and help make the world a better place!

Find a Person to Write To

You can write to people in government. Learn how to find your representatives at www.kidgovernor.org/student-action-resource-center/writing-to-your-elected-officials. The website also has a sample letter. Here are some people to look for:

- mayor or councilperson
- state senator or state representative
- members of the US Congress

Write and Send Your Letter

Here are some additional websites that have sample letters and can help you find out where to send your letters to:

- Open States, https://openstates.org
- United States House of Representatives, https://www.house.gov/representatives/find-your-representative
- United States Senate, https://www.senate.gov/senators/senators-contact.htm

Glossary

activist: a person who supports strong actions to stand up for a cause

climate change: a change in the usual weather conditions of a place over a long time

discrimination: treating people unfairly because of a factor such as race or religion

ecosystem: an interconnected community of plants, animals, weather, and other organisms

emission: a substance released or sent out into the atmosphere

evaporate: when a liquid turns into a gas

external migration: when people migrate from one country to another

fossil fuel: a fuel containing carbon that is formed from prehistoric animal and plant remains

migrant: a person who moves from one place to another

policy: a plan or rules that are used to guide actions in a government or organization

Learn More

Environmental Migration Facts for Kids
   https://kids.kiddle.co/Environmental_migration

Global Warming: Britannica Kids
   https://kids.britannica.com/students/article/global-warming/311438

Harts, Shannon H. *Climate Change and Earth's Population*. New York: PowerKids, 2022.

Kerry, Isaac. *Climate Change and Extreme Weather*. Minneapolis: Lerner Publications, 2023.

Loh-Hagan, Virginia. *Environmental Rights*. Ann Arbor, MI: Cherry Lake, 2021.

NASA Climate Kids: What Is Climate Change?
   https://climatekids.nasa.gov/climate-change-meaning/

Index

climate models, 6

fossil fuels, 6, 8, 19

global warming, 24–25, 28
greenhouse gases, 8, 14, 25

Ibrahim, Hindou Oumarou, 26

migration, 9–10, 12, 17, 21–24, 27–28

Pacific Climate Warriors, 19
Paris Agreement, 25

renewable energy, 8, 25

weather events, 12–13, 15–18

Photo Acknowledgments

Image credits: Mike Hollingshead/Getty Images, p. 5; Ashley Cooper/Getty Images, p. 6; Alexandros Michailidis/Shutterstock.com, p. 7; boscorelli/Shutterstock.com, p. 9; Salvacampillo/Shutterstock.com, p. 10; Warren Faidley/Getty Images, p. 11; Nelson Antoine/Shutterstock.com, p. 13; EFDN/Shutterstock.com, p. 15; FotoKina/Shutterstock.com, p. 16; Simone Torkington/EyeEm/Getty Images, p. 18; Danny Lehman/Getty Images, p. 21; RoschetzkyIstockPhoto/Getty Images, p. 22; solmaz daryani/Shutterstock.com, p. 23; Alexandros Michailidis/Shutterstock.com, p. 25; Fotoholica Press/LightRocket/Getty Images, p. 26; Ink Drop/Shutterstock.com, p. 27; Mimadeo/Shutterstock.com, p. 28.

Cover image: AP Photo/David J. Phillip.